Andrea Faussone

HELICAL COMPRESSION CYLINDRICAL SPRINGS

design, calculation and verification

Copyright © 2018 Andrea Faussone

All rights reserved.

UUID: 1a23f464-fd05-11e7-8743-17532927e555

This ebook was created with StreetLib Write
http://write.streetlib.com

Table of contents

1.0 PURPOSE	1
2.0 REFERENCES	2
3.0 APPLICABILITY	3
4.0 BASIC FORMULAS	4
5.0 MAXIMUM OPERATING STRESS	8
6.0 TANGENTIAL ELASTICITY MODULUS "G"	9
7.0 COILS	11
8.0 BLOCK LENGTH	13
9.0 MINIMUM INTER-COILS LENGTH S	14
10.0 WINDING DIAMETER	15
11.0 LATERAL STABILITY	17
12.0 TRANSVERSAL LOAD	21
13.0 NATURAL RESONANCE FREQUENCY	25
14.0 SCHEME OF CALCULATION PROCEDURE	27
ANNEX I	44

1.0 PURPOSE

The purpose of this publication is to provide the designer with a practical guide for the design of cylindrical compression springs.

The examples, which reflect the most usual cases, can be used as a design and verification check list. In this way, the product definition is obtained according to the boundary conditions and objectives.

2.0 REFERENCES

This document agrees with:

- **UNI - 7900** Part 2

3.0 APPLICABILITY

This document applies to cylindrical compression helix springs, of which active coils have a uniform inclination respect to the axis and which operate at a temperature not very different from the normal conditions (-30°C ÷ 80 °C).

It also applies to springs made either from wire or bar with a circular cross-section either from wire or bar with cross-rectangular section: the latter case is generally to avoid and must be adopted only when is not technically possible the use of wire or bar with a circular cross-section.

4.0 BASIC FORMULAS

4.1 Basic formulas for helical compression springs, made with wire or bar with a circular section of diameter "d"

4.1.2 torsional stress

The expression of the Torsional Stress T_k, measured in N/mm^2, generated by axial load **F** is the following:

$$T_k = \frac{8}{\pi} k \frac{D}{d^3} F$$

In which the expression of the measure unitless factor **k** of torsion stress correction is

$$k = \frac{4c-1}{4c-4} + \frac{0.615}{c}$$

This formula is valid for springs with:

- More than two active coils
- Winded with winding ratio **c=D/d** between **3** and **20**.

Values of the **k** factor as a function of the ratio **c** are set out in **Annex I**.

4.1.3 Torsion stress per unit of pressure that acts on the basis of the spring

The expression of the torsional stress per unit of pressure that acts on the base of the spring depends on the diameter chosen for the base surface. The base surface of the spring can be measured by the average diameter **D** or by the external diameter D_e or by the inside diameter D_i and takes the following forms:

$$w = \frac{\pi}{4} D^2 \frac{T_k}{F} = 2kc^3$$

$$w_e = \frac{\pi}{4} D_e^2 \frac{T_k}{F} = 2kc(c+1)^2$$

$$w_i = \frac{\pi}{4} D_i^2 \frac{T_k}{F} = 2kc(c-1)^2$$

Values of measure unitless **w**, **w $_e$** and **w $_i$**, are listed in the **Annex I** according to the winding ratio **c** and are used for the calculation of the springs as indicated below.

4.1.4 Displacement f caused by axial load F

The displacement expression **f**, measured in mm, caused by axial load **F** is the following:

$$f = \frac{8}{G} \frac{D^3}{d^4} iF$$

For the values to be assigned to the tangential elasticity modulus G of the material used, see § 6.0. For the values to be assigned to the number of active coils i, see § 7.2.

4.2 Basic formulas for cylindrical helix springs operating in compression, made with wire or bars with rectangular cross-section with radial width "l" and height "h"

4.2.2 torsional stress

The expression of the torsion stress T_k, measured in N/mm^2, caused by axial load F takes the following appearance

$$T_k = \beta \frac{D}{\sqrt{l^3 h^3}} F$$

4.2.2.1 Correction factor "β" of the torsion stress

The correction factor **β** of the torsional stress is a function of the cross-section shape ratio *m=l/h* and of the winding ratio **c**. **Diagram 1** shows the values to assign to the correction factor **β** according to the cross-section shape ratio *m* for different values of the winding ratio *c* between 3 and ∞ in the case of a rectangular cross-section of the wire or bar.

The validity of the diagram is limited to the case of springs having **more than two active coils**.

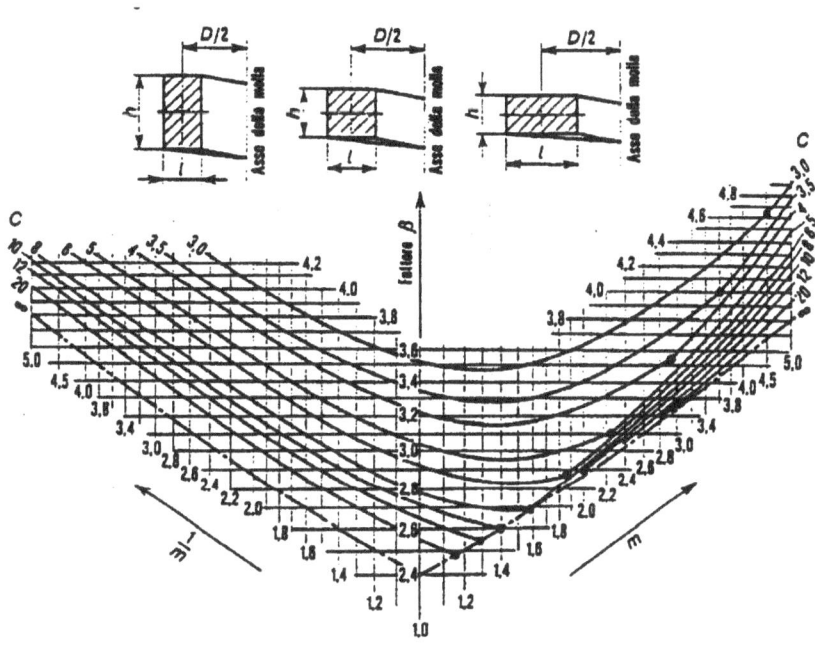

Diagram 1 - Values of correction factor β in function of cross-section shape ratio m and of winding ratio c

4.2.3 Displacement f caused by axial load F

The displacement expression **f**, measured in mm, caused by axial load **F** is the following

$$f = \frac{\varepsilon}{G} \frac{D^3}{l^2 h^2} iF$$

For the values to assign to the tangential elasticity modulus **G** of the used material, see § 6.0. For the values to assign to the number of active coils **i**, see § 7.2.

4.2.3.1 Displacement correction factor ε

The displacement correction factor **ε** in the displacement formula is function of the cross-section shape factor **m**. In the following table there are the values for the displacement correction factor **ε** in function of the cross-section shape factor **m** between 0.2 and 5.0 for the rectangular cross-section wire or bar.

Table I - Displacement correction factor ε

m	0.2	0.4	0.6	0.8	1.0	1.2	1.4	1.6	1.8
ε	13.48	7.87	6.25	5.71	5.59	5.67	5.88	6.17	6.50
M	2.0	2.2	2.4	2.6	2.8	3.0	3.2	3.4	3.6
ε	6.87	7.26	7.67	8.09	8.51	8.95	9.39	9.83	10.28
m	3.8	4.0	4.2	4.4	4.6	4.8	5.0		
ε	10.73	11.19	11.63	12.11	12.50	12.92	13.48		

5.0 MAXIMUM OPERATING STRESS

5.1 Static stress condition

The springs operating under constant load or subjected to occasional load variations, so diluted in time being less than **10,000** alternations throughout the entire life of the spring. For springs subjected to static stress condition, yielding or breakage may occur only as a result of the elastic limit of the material and the maximum stress allowable is caused by the characteristics of the same resistance and is fixed as a fraction of the tensile breaking strength **R**.

For cold-winded helical compression cylindrical springs , **without hardening treatment**, the ratio between the highest torsion stress achievable working T_{kn} and the tensile strength **R** of the material used, shall not exceed the percentages following:

- 55% for unalloyed or alloyed steel wires or bars, pre-filled;
- 50% for unalloyed or alloyed steel wires or bars, patented;
- 50% for stainless steel or strongly alloyed steel wires and bars;
- 40% for phosphor bronze or silicon bronze wires, drawn by springs;
- 40% for hard brass wires, drawn by springs.

For hot-winded or cold-winded cylindrical compression coil springs and **subsequently hardened**, the ratio, between the highest torsion stress achievable working T_{kn} and the tensile breaking load **R** of the material used, must be contained within the maximum limit of 55% for **unalloyed or alloyed steel wires or bars to harden**.

5.2 Dynamic stress

The springs operating under a periodically variable load are considered to be subjected to dynamic tension, between two fixed values or subjected to variable loads in an occasional manner, with a frequency such as to **totalize at least 10,000** alternations throughout the entire life of the spring. The breakage of the springs working under dynamic stress is caused by fatigue, after a number of alternating stresses as well as by the maximum working stress, also by the extension of the stress field in which the spring operates.

The maximum permissible working values for the correct torsional stresses of the helix cylindrical compression springs under dynamic stress comes from the diagrams of **Woehler** or **Goodman** of the material that you intend to use. These diagrams are available from the steel mills, at least for the common material types used for the construction of the springs.

The state of the surface, especially at the inner cylindrical cloak, is essential for the fatigue strength of a spring. The surface must be smooth, **without signs or incisions** and t he **superficial decarbonization** must be as far as possible **eliminated** .

The **peening** with metallic grit, to equal to other conditions, increases the fatigue strength of a spring under a dynamic stress.

6.0 TANGENTIAL ELASTICITY MODULUS "G"

6.1 Influential characteristics

The value of the tangential elasticity modulus **G** depends not only on the type of material, but also on the working temperature, on the intensity of any settling operation and, in the case of steel, on the presence of surface decarburization.

6.1.2 Influence of temperature on the tangential elasticity modulus G

The influence of the **temperature t** of the environment in which the spring works is represented by the formula:

$$G = G_{20}[1 - r(t - 20)]$$

Where **G** and **G $_{20}$**, both measured in N/mm^2, represent the tangential elasticity modulus of the material, respectively at temperatures **t** and **20°C**, and **r** is the coefficient of which value is:

- 0.25E-3 for unalloyed or alloyed steels;
- 0.40E-3 for stainless steels;
- 0.40E-3 for phosphor bronze and hard brass.

The values indicated in § 6.2 refer to the working temperature of 20°C, but taking into account the values of the temperature, the coefficient **r**, indicated in this point, can be used in the working temperature range from 0 to 40° C, without correction. They can be used immediately, being the reduction for a moderate adjustment action already foreseen and, in the case of hot-wrapped steel springs, the reduction due to the presence of a modest superficial decarburization, contained within the limits considered normally acceptable.

The operation of **adjustment** determines, all other conditions being equal, a reduction in the value of the tangential elasticity modulus of steels, the more sensitive the more intense the operation was.

The presence of the superficial decarbonization produces, all other conditions being equal, a decrease in the tangential elasticity of the steels, the more relevant the greater the degree and thickness of decarburization are.

Due to the intrinsic characteristics of their production process, the hot-winded springs are, far more than those cold-winded, susceptible to the phenomenon of surface decarburization.

6.2 Values of G at 20°C

The values that we recommend to assign to the tangential elasticity module G20 of the materials most used in the construction of helical compression cylindrical springs, measured in N/mm^2, are the following:

- 78,500 ÷ 81,500 for unalloyed or alloyed steel, for cold-winded springs without subsequent hardening treatment;
- 75,500 ÷ 78,500 for steel for hot-winded or cold-winded springs and subsequently reclaimed;
- 71'500 ÷ 75'500 for stainless steel for springs;
- 41'000 ÷ 43'000 for silicon bronze and phosphor bronze for springs;
- 34'000 ÷ 36'000 for the hard drawn brass for springs.

The values indicated above are subject to variation in relation to the data supplied by the manufacturers of the spring materials.

7.0 COILS

7.1 General criteria

It is convenient, to obtain a correct axial thrust, that the total number of coils i_t is an odd multiple integer of half coil or at least in case it is less than 8, an odd multiple of coil quarter.

The use of a integer of coils is to be discouraged in the case of springs working in dynamic stress conditions.

7.2 Active coils

The number of working coils **i**, which appears in the formulas in § 4.1.4 and § 4.2.3, is related to the total number of coils i_t from the formula:

$$i = i_t - i_m$$

Where i_m represents the number of inactive coils, that do not take part in the process of elastic deformation.

7.3 Inactive coils

The value to be assigned to the number of inactive coils i_m depends on the configuration of the spring ends

- i_m equal to 0: in the springs with the two ends open (see figure 2).
- i_m can vary from 1.25 to 1.75: in the springs with the two open and ground ends, or closed and ground ends, or closed (see Figure 2) according to the actual angular development of the end support.
- i_m takes the value of 1.50: in the most frequent case in which this development is 270°.
- i_m is equal to 2: in the springs with the two closed ends ground, or closed, or tapered closed and ground (see Figure 2), unless the contact does not extend for a certain development, in which case i_m is equal to 2 plus the coils number corresponding to the developed length of coils, measured on the two ends.

Type end	closed	open	closed, ground	open, ground	Tapered, closed, ground
	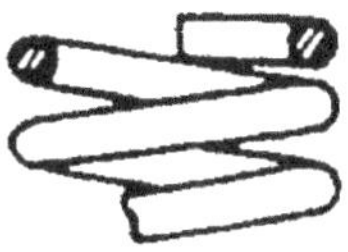	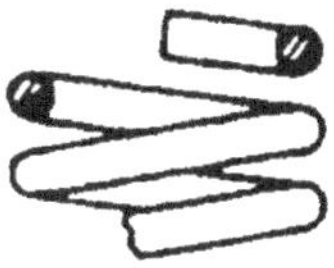			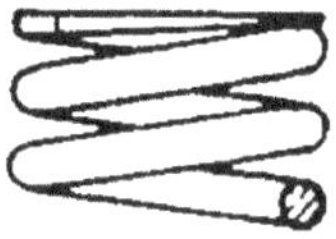
im	2	0	2	1.5	2
L_b	(it+1)d		(it-0.5)d		

Figure-2 - number of inactive coils according to the end type

8.0 BLOCK LENGTH

8.1 Influential characteristics

The block length l_b depends, as well as on the total number of coils i_t and the diameter of the wire or bar **d** (or from height **h** in the case of springs constructed with wire or bar with rectangular cross-section), also to the ends shape and the protective coating thickness. In calculating its maximum value, the dimensional tolerance, the circularity tolerance of the diameter **d** of the wire or bar, the number tolerance of coils i_t, the coating thickness and any ends execution requirements must be taken into account.

8.1.2 Ends not ground and not tapered

In the case of springs with not ground ends, not tapered, whether open or closed, the block length, measured in mm, is expressed by one of the two formulas:

$$L_b = (i_t + 1)d$$

$$L_b = (i_t + 1)h$$

valid, respectively, for the springs with circular cross-section and rectangular cross-section of the wire or bar.

8.1.3 Only ground or ground and tapered ends

In the case of springs with only ground or ground and tapered ends, the block length, measured in mm, is expressed by one of the two formulas:

$$L_b = (i_t - 0.5)d$$

$$L_b = (i_t - 0.5)h$$

valid, respectively, for the springs with circular cross-section and rectangular cross-section of the wire or bar.

9.0 MINIMUM INTER-COILS LENGTH S

The minimum inter-coils length S, measured in mm, is the sum of each inter-coil length, which is advisable to remain available at the maximum working load of the spring. It can be calculated using one of the two formulas:

$$S = s(d + 0.5)$$

$$S = s(h + 0.5)$$

valid respectively for the springs with circular cross-section and rectangular cross-section of the wire or bar, with **d** or **h** not less than 1.0 mm.

The measure unitless coefficient **s**, which appears in the previous formulas, is a function of the winding ratio **c**.

The values to be assigned to **s**, in correspondence with winding ratios **c** between 3 and 15, are shown in the following table.

Table II – Values of the coefficient s

c	3	4	5	6	7	8	9	10	11	12	13	14	15
s	0.091	0.094	0.100	0.109	0.121	0.136	0.154	0.175	0.199	0.226	0.256	0.289	0.325

10.0 WINDING DIAMETER

10.1 Spring expansion with fixed ends

In the case of springs with a circular cross-section of the wire or bar and with both ends placed in the impossibility of unwinding, the application of the load causes a reduction in the coils position angle and, consequently, an increase in the mean, external and internal winding diameters.

The increase ▲D, measured in mm, of the mean winding diameter, when the spring passes from the load-free condition to the blocking position, is expressed by the formula:

$$\Delta D = \frac{p_0^2 - d^2}{20D}$$

The p_0 is the pitch, measured in mm, of the active coils of the loadless spring expressed by the formula:

$$p_0 = \frac{L_0 - L_b}{i} + d$$

10.2 Spring expansion with free ends

In the case of springs with a circular cross-section of the wire or bar and with at least one of the ends completely free to unwind without friction, the application of the load causes, in addition to the phenomenon described in §10.1, the unwinding of the compressed coils and therefore a further increase in the mean, external and internal winding diameters of the active coils.

The increase ▲D, measured in mm, of the mean winding diameter, when the spring passes from the loadless condition to the blocking position, is expressed by the formula:

$$\Delta D = \frac{p_0^2 - 0.8 p_0 d - 0.2 d^2}{10D}$$

The p_0 is the pitch, measured in mm, of the active coils of the loadless spring expressed by the formula § 10.1.

10.3 Springs with wire or bar with circular cross-section

In practice, the constraint conditions of the ends of the spring rarely correspond to the limit case illustrated in § 10.1. From the observations made, the formula set out in § 10.2 is practically usable in the majority of cases.

10.4 Springs with wire or bar with rectangular cross-section

When the springs are made by rectangular cross-section wire or bar, the phenomenon is more complex and, due to the application of the load, a deformation of the coils can even occur, with a consequent reduction of the winding diameter. The variation of the winding diameter must be detected experimentally, case by case.

11.0 LATERAL STABILITY

11.1 Spring behavior

When certain conditions are achived, the helical cylindrical springs, subjected to axial compression load, tend to bend laterally, behaving in the same way as a tip-loaded rod. Designing, it may be necessary to verify that, under no working conditions, the lateral stability of the spring is lost.

The lateral stability of a spring is strictly connected as well as to the fixing conditions of the ends, even to the values assumed by the **degree of slenderness** $\lambda = L_0/D$ and to the **inflection ratio** $\varphi - F/L_0$ of the spring.

11.2 Critical inflection ratio

For each spring with defined geometrical characteristics there is in practice a critical value of the deflection ratio φ_{crit} above which the spring is devoid of lateral stability.

This measure unitless value can be calculated using the formula:

$$\varphi_{crit} = X \left[1 - \sqrt{1 - Y(Z/\lambda)^2} \right]$$

Where **X**, **Y** *and* **Z** are three measure unitless coefficients linked to the spring geometrical characteristics, to the material used and to the ends constraint conditions.

The value of the coefficient **X**, which appears in the formula, can be calculated with one of the two formulas:

$$X = \frac{0.50}{1 - \dfrac{G}{E}}$$

$$X = \frac{0.50}{1 - \dfrac{4.71\,G}{m\varepsilon\,E}}$$

The first formula is valid for springs made with wire or bar with circular cross-section. The second formula is valid for springs made with wire or bar with rectangular cross-section, which have the shape ratio **m** of the section and the correction factor **ε** of the displacement: for the values of **ε** please refer to § 4.2.3.1.

The value of the coefficient Y, which appears in the formula, can be calculated with one of the two formulas:

$$y = 19.74 \frac{1 - \dfrac{G}{E}}{1 + 2\dfrac{G}{E}}$$

$$Y = 19.74 \frac{1 - \dfrac{4.71}{m\varepsilon}\dfrac{G}{E}}{1 + 9.42\dfrac{m}{\varepsilon}\dfrac{G}{E}}$$

The first formula is valid for springs made with wire or bar with circular cross-section. The second formula is valid for springs made with wire or bar with rectangular cross-section, which have the shape ratio m of the section and the correction factor ε of the displacement : for the values of ε please refer to § 4.2.3.1

The ratio between the tangential elasticity modulus G and the modulus of normal elasticity E, which appears in the formulas in § 11.2, has the following mean values:

- 0.385 for steel
- 0.420 for bronze
- 0.360 for brass

The coefficient Z, which appears in the formula, depends exclusively on the constraint conditions of the ends of the spring.

The following values can be assigned to it (see figure 3):

- 2.0 if both end plates are interlocked and laterally guided in their relative motion;
- 1.5 if one of the end plates is interlocked and the other is hinged and their relative motion is guided laterally;
- 1.0 if both end plates are hinged and laterally guided in their relative motion;
- 1.0 if both spring ends are interlocked, but not laterally guided in their relative motion;
- 0.5 if one of the end plates is interlocked and the other is free.

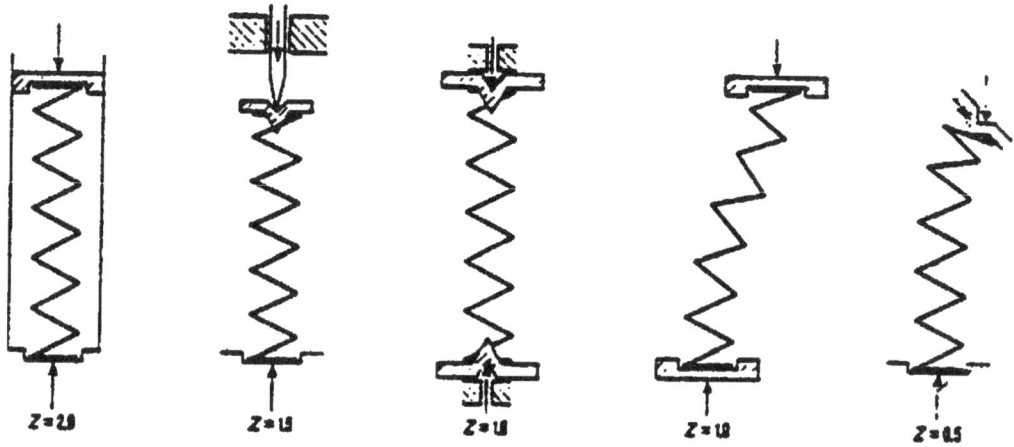

Figure-3 the Z coefficient in different constraint conditions of the ends

11.3 Evaluation of lateral stability

The values of the ratio Z/λ appear in **Table III** calculated with the formula in § 11.2, in correspondence with prefixed values of the critical deflection ratio φ_{crit} for springs made of steel, either with wire or bar with a circular cross-section, or with wire or bar with rectangular cross-section with different shape ratios m of the section.

Figure 4 shows the ratio Z/λ **curses** in function of the critical deflection ratio φ_{crit} for spring made of wire or bar with circular cross-section, most frequent case, and of wire bar with **square cross-section, less frequent case.** On the the diagram, the curves separate the area in which springs have lateral stability from that in which they have not.

To be able to foresee drawing the behavior of a steel spring, made with wire or bar with a circular or square cross-section, taking advantage of the diagram shown in Figure 4, it is sufficient to calculate the ratio Z/λ and verify that the deflection ratio variations corresponding to the working conditions are entirely within the stability zone.

Table III - Ratio Z/λ

Section		φ_{crit}										
		0.05	0.1	0.2	0.3	0.4	0.5	0.6	0.7	0.8	0.9	1.0
		Z/λ										
CIRCULAR		0.132	0.183	0.251	0.296	0.329	0.352	0.368	0.378	0.382	0.380	0.372
Rectangular with shape ratio m	0.25	0.103	0.144	0.200	0.240	0.272	0.297	0.320	0.339	0.354	0.367	0.378
	0.33	0.106	0.149	0.206	0.246	0.278	0.303	0.324	0.340	0.356	0.368	0.372
	0.5	0.112	0.156	0.215	0.256	0.288	0.313	0.332	0.346	0.357	0.364	0.368
	0.66	0.118	0.164	0.225	0.267	0.298	0.321	0.338	0.350	0.357	0.360	0.358
	1	0.127	0.177	0.240	0.283	0.312	0.333	0.345	0.351	0.351	0.344	0.330
	1.5	0.136	0.189	0.255	0.297	0.324	0.341	0.347	0.345	0.334	0.313	0.279
	2	0.141	0.195	0.262	0.304	0.330	0.343	0.346	0.338	0.319	0.286	0.234
	3	0.146	0.202	0.270	0.312	0.336	0.346	0.346	0.330	0.302	0.255	0.174
	5	0.150	0.207	0.277	0.318	0.341	0.349	0.345	0.326	0.290	0.231	0.114

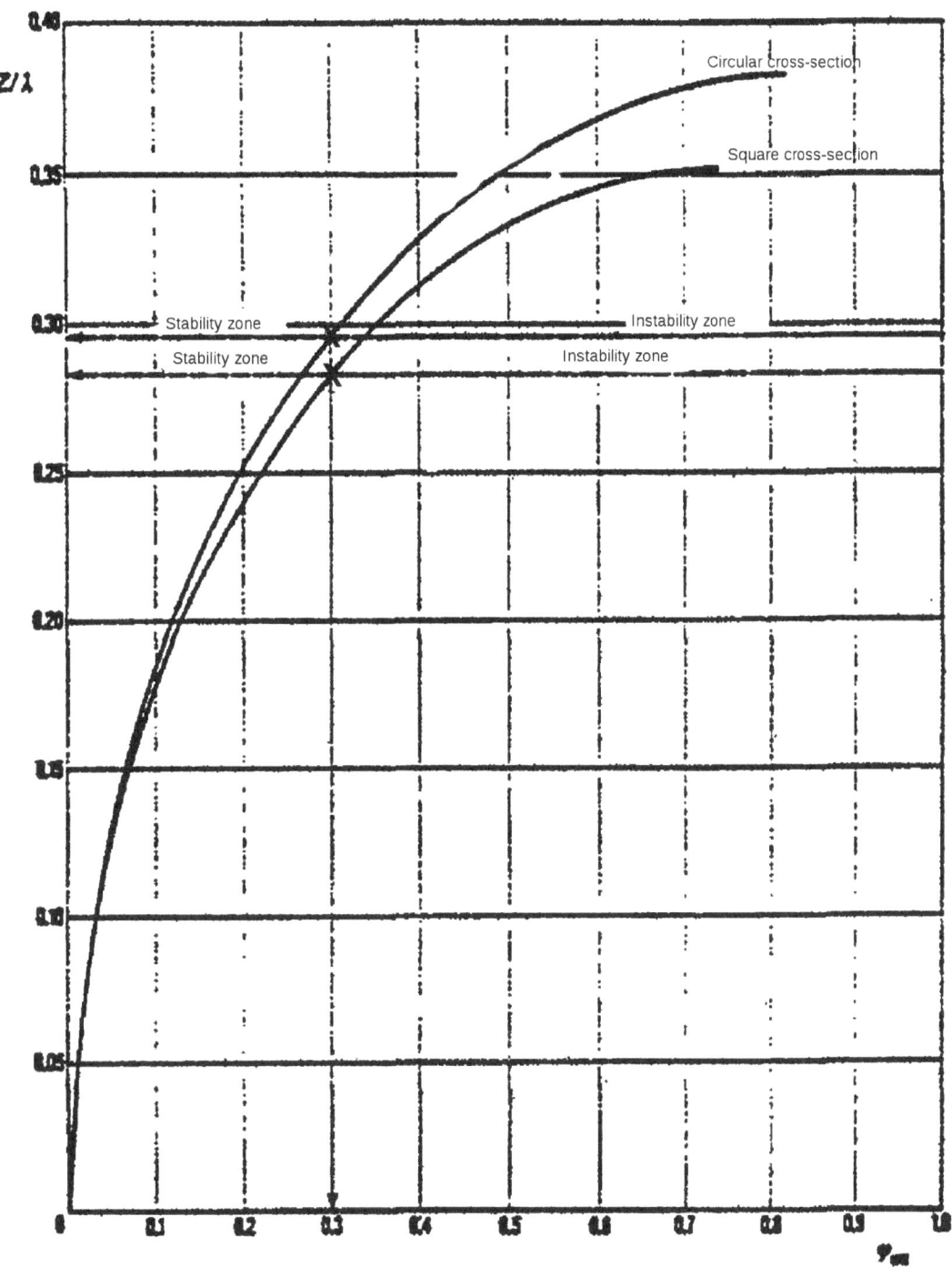

Figure-4 lateral stability zone of the springs

12.0 TRANSVERSAL LOAD

12.1 Spring behavior

A spring, interlocked at one end and free from the other to perform translation movements in a plane containing the axis of the helix, subjected to axial load **F** and simultaneously to transverse load F_t, contained in the translation plane, undergoes an axial displacement **f** and a lateral displacement **t**.

12.2 Axial displacement

The axial displacement **f** can be calculated using the formula given in § 4.1.4 or t by the formula in § 4.2.3, according to whether the section of the wire or bar, with which the spring is constructed, has a circular or rectangular cross-section.

12.3 Lateral displacement

The lateral displacement **t**, measured in mm, can be calculated using the formula:

$$t = L\frac{F_t}{F}\left\{(1+\xi\frac{F}{L})\frac{tg(\frac{1}{2}qL)}{\frac{1}{2}qL} - 1\right\}$$

Where **L** is the axial length assumed by the spring under the action of the load **F**.
The physical dimension **q**, measured in mm^{-1}, which appears in the formula of the lateral displacement, is the warping factor of the spring and is calculated using the formula:

$$q = \sqrt{\psi\frac{F}{L}(1+\xi\frac{F}{L})}$$

The physical dimension ξ, measured in mm/N, which appears in both expressions of the lateral displacement and in the warping factor expression, can be calculated using one of the two formulas:

$$\xi = 8\frac{iD^3}{Ed^4}$$

$$\xi = \frac{4.71}{m} \frac{iD^3}{El^2h^2}$$

The first formula is valid for springs made with wire or bar with circular cross-section. The second formula is valid for springs made with wire or bars with rectangular cross-section, having a shape ratio **m** of the section.

The physical dimension **ψ**, expressed in mm $^{-1}$ N $^{-1}$, which appears in the expression of the spring warping factor, can be calculated using one of the two formulas:

$$\psi = 32(1 + 0.5\frac{E}{G})\frac{iD}{Ed^4}$$

$$\psi = 18.45m(1 + \frac{\varepsilon}{9.42m}\frac{E}{G})\frac{iD}{El^2h^2}$$

The first formula is valid for springs made with wire or bar with circular cross-section. The second formula is used in the case of springs made with wire or bar with rectangular cross-section, which has the shape ratio **m** of the section and the displacement correction factor **ε** : for the values of the latter refer to § 4.2.3.1 and to Table I.

12.4 Lateral displacement with zero compression load

In case the axial load **F** is null and the spring is subjected only to transverse load **F t**, the expression of the lateral displacement **t $_0$**, measured in mm, assumes the following simplified form:

$$t_0 = F_t(\xi + \frac{1}{12}L_0^2\psi)$$

Where L $_0$ represents the axial length of the spring in an axial load-free configuration. ξ and ψ assume the values pressed by the formulas in § 12.3.

12.5 Torsion stress in the presence of a traversal load

The presence of the traversal load **F $_t$** modifies the distribution of the torsional stresses, which are no longer uniform and quantitatively equal for all the cross sections of the wire or bar, with which the spring is made, but the torsional stresses have an absolute maximum evaluable, measured in N/mm^2, with the formula:

$$T_{max} = T \cdot T_k$$

This maximum is achieved in the cross sections of the active coils of the spring, which are located on the traversal load plan.

The physical dimension T_k is the torsional stress caused by the axial compression load only **F**. It can be calculated using the formula given in § 4.1.2, or using the formula given in § 4.2.2, according to whether the cross-section of the wire or bar with which the spring is built is circular or rectangular.

The factor **T**, which appears in the expression is a measure unitless corrective factor, which can be evaluated using the formula:

$$T = 1 + \frac{t}{D} + \frac{LF_t}{DF}$$

The formula is applicable only in the case of axial compression load **F** different from zero and lateral displacement **t**, which appears in it, must be calculated using the formula shown in § 12.3.

12.6 Torsion stress with no compression axial load

In case in which the axial compression load **F** is null and the spring is subjected to the traversal load only F_t, The torsion stress assumes its absolute maximum value always in the cross sections of the active end coils of the spring, which are located in the axial plane of the load position F_t. This maximum value, measured in N/mm^2, can be expressed using one of the two formulas:

$$T_{max} = \frac{8}{\pi} k \frac{L_0}{d^3} F_t$$

$$T_{max} = \beta \frac{L_0}{\sqrt{l^3 h^3}} F_t$$

The first of the two formulas is valid in the case of springs made with wire or bar with a circular cross-section. The term k is the correction factor of the torsion stress, the expression of which is given in § 4.1.2 and values are tabulated in **Annex** I.

The second of the two formulas is applied in the case of springs made with wire or bar with rectangular cross-section. The factor **β** is the **correction factor of the torsion stress**, values of which can be obtained from the diagram shown in **diagram 1**.

13.0 NATURAL RESONANCE FREQUENCY

13.1 Spring with two interlocked ends

An helical cylindrical spring, having the two ends resting on two rigid plates, has a natural resonance frequency **N** measured in s^{-1}, evaluable by one of the two formulas.

$$N = K \frac{d}{iD^2}$$

$$N = K' \frac{\sqrt{lh}}{iD^2}$$

The first formula is valid for springs made with wire or bar with circular cross-section. The second one is applied in the case of springs made of rectangular cross-section wire or bar.

If the natural resonance frequency is measured in s^{-1}, the two coefficients **K** and **K'** assume the following expressions:

$$K = 3560 \sqrt{\frac{G}{\rho}}$$

$$K' = \frac{8920}{\sqrt{\varepsilon}} \sqrt{\frac{G}{\rho}}$$

In the expressions above **G** is the **tangential elasticity modulus**, measured in N/mm^2, ρ is the **volumetric mass**, measured in kg/dm^3, and ε is the **correction factor of the displacement**, valid for springs made with wire or bar with rectangular cross-section, values of which are shown in **Table I**.

In the particular case of springs made of pre-hardened steel wire for cold-winded springs (such as the return springs for the distribution of the driving and rapid piston-driven machines), it can be set G=81'500 N/mm^2 and ρ= 7.85 kg/dm^3, for which the formulas indicated become:

$$K = 3.63E5$$

$$K' = \frac{9.09E05}{\sqrt{\varepsilon}}$$

13.2 Springs with only one end interlocked

If the spring has only one interlocked end and the other is free, the natural resonance frequency is half that of the same spring with both ends blocked.

13.3 Springs operating under dynamic loads

For springs working under periodically oscillating loads (such as the return springs for the distribution of the driving and operating machines with piston) it is necessary to check, during the drawing, that the natural frequency, calculated with the formulas in § 13.1, is more high of the highest harmonic of the frequency of exercise still susceptible, for its intensity, to trigger the phenomenon of entry into resonance.

14.0 SCHEME OF CALCULATION PROCEDURE

14.1 Calculation of a helical compression cylindrical spring, to be realised with wire or bar with a circular cross-section working in static stress condition

Starting data are, as a rule, the following:

- One of the two winding diameters D_e or D_i with relative tolerance (less frequent is the assignment of the mean winding diameter D);
- The operating load F and the corresponding displacement f or, if the load is variable (with the limitations indicated in § 5.1), the maximum working load F_n and the corresponding displacement f_n.
- Possible particular working conditions (for example: working environment temperature that exceeds normal limits).
- The configuration to be assigned to the spring ends is also established during the drawing.

In the case in which it is renounced to accurately prefix the outer winding diameter as a starting point D_e, or the internal diameter D_i, the calculation can be set as indicated in the alternative scheme shown in § 14.4.

1. On the basis of the general overall dimensions available and the working conditions of the spring, it is envisaged whether it must be cold or hot winded and the type of material it is produced and selected.

2. Consequently, according to the prescription given in § 5.1, if the spring must be cold winded, or hot winded, the value of the torsion stress is established. T_k corresponding to load F (respectively T_{kn} to F_n, if the load is variable).

3. According to the external diameter has been assigned D_e or the internal diameter D_i or the mean winding diameter D, the **torsional stress per unit of pressure** exerted on the basis of the spring is calculated using the corresponding formula in § 4.1.3 W_e or W_i or W.

4. The spring winding ratio c is then evaluated using **Annex I**. Finally, the value, measured in mm, of the wire or bar diameter d is obtained using one of the three formulas:

$$d = \frac{D_e}{c+1}$$

$$d = \frac{D_i}{c-1}$$

$$d = \frac{D}{c}$$

5. The diameter of the wire or bar is then rounded to the nearest unified value and, based on this value, the mean winding diameter is calculated, in the case, usually more frequently, that the external or the internal diameter has been assigned. The calculation formulas are respectively:

$$D = D_e - d$$

$$D = D_i + d$$

- 6. From the formula in § 4.1.4 we obtain the explicit expression of the number of active coils:

$$i = \frac{G}{8} \frac{d^4}{D^3} \frac{f}{F}$$

of which value can be calculated immediately, in fact the terms appearing in the second member of the expression are known.

7. Knowing the number of active coils, we then go back to the total number of spring coils i_t, taking into account the configuration of the ends and evaluating the number of inactive coils on the basis of what is indicated in § 7.3.

8. The value of i_t calculated is then rounded on the basis of the specifications contained in § 7.1 and, of course, the same rounding is then made to the number of active coils.

9. As the calculated values of **d** and of **i** are rounded according to point 4.5 and 6, a verification calculation should be carried out before proceeding further.

    a. First of all, the winding ratio **c** is evaluated and from this, through the expression in § 4.1.2, the actual value of the factor **k** of correction of the torsion stress is obtained.

b. Thus, using the expression in §4.1.2, in which all the terms appearing in the second member are known, the correct torsional stress T_k corresponding to the working load F is calculated.

c. Finally, the expression in § 4.1.4, in which all the terms are known, *allows checking that the spring elastic characteristics correspond to the design data.*

If the spring verification calculation provides unsatisfactory results, the calculation is repeated, suitably modifying the rounding of d and of i and, if necessary and the configuration of the terminals makes it possible, by changing the number of coils.

10. Using the formulas given in § 8.0, the block length of the spring L_b is evaluated and the minimum inter-coils length $S.^1$

11. The length L of the spring at the load F, measured in mm, is then calculated using the following formulas and the length L_0 of the spring in the absence of a load, measured in mm, as well as the blocking displacement f_b, measured in mm:

$$L = L_b + S$$

$$L_0 = L_b + S + f$$

$$f_b = L_0 - L_b$$

Reasons for use may impose the increase of the minimum inter-coils length beyond the value resulting from the application of the formulas given in § 8.0. In this case, also the length L and L_0 are increased to the same extent.

12. The theoretical load F_b supported by the spring in the blocked position is calculated, using the proportion:

$$F_b : F = f_b : f$$

being, respectively, f_b and f the displacements in the blocked condition and under the load F.

13. Then, using the expression in § 4.1.2, the torsion stress T_{kb} is evaluated.

14. It is assigned to high test stress T_{kc} a value greater than at least 10% of the maximum working stress T_k, but, in any case, less than the elastic torsional limit of the used material. The spring test load F_c is then calculated F_c, using the proportion:

$$F_c : F = T_{kc} : T_k$$

May occur that the test load F_c is greater than the block load F_b; in this case the test load calculated is not achievable and we adopt the spring blocking load as the test load.

15. Finally, the expansion of the winding diameter is calculated with the formulas given in § 10.0 and the lateral stability of the spring is checked with the procedure indicated in § 11.0.

14.2 *Calculation of a helical compression cylindrical spring, to be realized with wire or bar with a circular cross-section working in dynamic stress condition*

the known data are, in the most common case, the following:

- One of the two winding diameters D_e or D_i, with relative tolerance: if necessary, the mean diameter D can be assigned instead of them;
- The minimum value F_1 and the maximum value F_n of the working load and the spring stroke between the two loads ▲ L ;
- the maximum frequency of periodic oscillation of the working spring N_0;
- Possible particular working conditions (for example: working environment temperature exeeding normal limits).
- The shape of the ends is also defined during the drawing.

1. Depending on the geometrical and working conditions, it is established whether the spring must be cold or hot winded and the type of material to be used is chosen.

2. Keeping in mind that working stress T_{kl} and T_{kn} are proportional to the corresponding loads F_1 and F_n, the values are fixed on the basis of the fatigue strength of the selected material and of any additional processing procedures (for example **peening**) planned to increase these capacities.

3. With the same methods as shown for the case of the spring operating under static conditions, the diameter of the wire or of the bar and the mean winding diameter (if this has not already been provided as a starting data) are calculated using the procedure described in § 14.1. points 4 and 5.

4. The explicit expression of the number of active coils is immediately deduced from the formula in § 4.1.4 and applies:

$$i = \frac{G}{8} \frac{d^4}{D^3} \frac{\Delta L}{F_n - F_1}$$

5. The calculation of the total coils number of the spring, its rounding and the consequent rounding of the number of active coils, is performed in a manner similar to that specified for the case of the spring operating under static stress in § 14.1 point 6 and following.

6. Before proceeding further, the calculation of the verification of the torsion stress corresponding to the working loads is carried out and is checked that the elastic characteristics of the spring correspond to the values required in the drawing.

7. The spring block length L_b and the minimum inter-coils length S, according to the

provisions contained in § 8.0, are evaluated.

8. The lengths L_n and L_1 of the spring are then calculated under the load F_n and the load F_1, measured in mm, by the formulas:

$$L_n = L_b + S$$

$$L_1 = L_b + S + \Delta L$$

9. The proportionality between loads and displacements allows to obtain the values of the latter, measured in mm, at the loads F_n and F_1

$$f_n = \Delta L \frac{F_n}{F_n - F_1}$$

$$f_1 = \Delta L \frac{F_1}{F_n - F_1}$$

10. It is therefore possible to calculate length L_0 of the spring in the absence of load and the block displacement f_b, both measured in mm, by the formulas:

$$L_0 = L_b + S + f_n$$

$$f_b = L_0 - L_b$$

11. Naturally, also in this case, as in that illustrated in § 14.1 point 11, functional reasons may impose a increment of the three lengths, of null load and the corresponding working conditions due to the loads F_1 and F_n; so the values L_n, L_1 and L_0 can be increased to the same extent with respect to the values resulting from the formulas.

12. The theoretical load F_b supported by the spring in the blocked position is calculated, using the proportion:

$$F_b : F_n = f_b : f_n$$

being, respectively, f_b and f_n, the displacement in the block position and under the load F_n.

13. Then, using the expression in § 4.1.2, the torsion stress T_{kb} is evaluated.

14. It is assigned to high test stress T_{kc} a value greater than at least 10% of the maximum working stress T_k, but, in any case, less than the elastic torsional limit of the used material. The spring load test F_c is then calculated, using the proportion:

$$F_c : F_n = f_c : f_n$$

In the event that the load test is greater than the block load F_b, the blocking load of the spring is used as the test load.

15. The expansion of the winding diameter is calculated using the formulas indicated in § 10.0.

16. The lateral stability of the spring is checked with the procedure indicated in § 11.0

17. We verify that the natural resonance spring frequency is higher than the highest harmonic of the working frequency by using the formula given in § 13.0.

14.3 Calculation of an helical compression cylindrical spring, to be realised with a rectangular cross-section wire or bar

The calculation of a helical compression cylindrical spring, to make with wire or bar with rectangular cross-section, is carried out by a procedure similar to that indicated in § 14.1 and § 14.2. The greater complexity of the formulas and the presence of the variable additional, represented by the shape ratio **m** of the rectangular cross-section, are compensated by the fact that some unknown quantities already know the order of magnitude, based on the preventive calculation carried out to verify the possibility of making the spring with wire or bar with a circular cross-section.

14.4 Scheme of an alternative calculation procedure compared to that reported in § 14

As an alternative to the calculation procedure shown in § 14, if we renounce accurately to prefix the outer winding diameter D_e, or the inside diameter D_i, we can set the calculation, as indicated below, placing only a maximum limit to the value of D_e, or a minimum limit to the value of D_i and instead of imposing, as a datum of the problem, one of the following three dimensions:

- spring length under an assigned load
- length of development of the wire or bar with which the spring is made
- spring mass (or rather, the mass of the wire or bar constituting the spring **before ends grinding**).

The calculation procedure only concerns the compression springs of the following two types:

- with closed and ground ends (see figure 2)
- with closed, ground and tapered ends (see figure 2)

14.4.2 Diagram of calculation procedure in the case in which the length of the spring working with assigned load is prefixed

The starting data available are the following:

- The maximum value of the outer winding diameter D_{emax} or, alternatively, the minimum value of the internal winding diameter D_{imin};
- Minimum and maximum working loads F_1 and F_n;
- The corresponding variation in length ▲ L (or the corresponding displacement variation ▲ f being: $\Delta L = \Delta f$);
- the length L_n under the load F_n (or the length L_1 under the load F_1; in this case we would find immediately L_n, being: $L_n = L_1 - \blacktriangle L$);
- the torsion stress T_{kn}, at the maximum working load F_n;
- the tangential elasticity modulus G of the material to be used for the construction of the spring
- It is also established, at this stage of the calculation, that the spring is constructed with a minimum inter-coils length S that for simplicity is better expressing with the formula:

$$S = s \, i \, d$$

instead of the one indicated in § 9.0.

To the coefficient **s** the value assigned is:

$$s = 0.01 \, (0.15 c^2 - 0.75 c + 10)$$

1. According to the data we have, the maximum value of the external winding diameter D_{emax} or the minimum value of the internal winding diameter D_{imin}, we calculate the corresponding torsion stress per unit of pressure exerted on the spring base using one of the formulas (see point § 4.1.3):

$$w_e = \frac{\pi}{4} D_{e\max}^2 \frac{T_{kn}}{F_n}$$

$$w_i = \frac{\pi}{4} D_{i\min}^2 \frac{T_{kn}}{F_n}$$

Using the **annex I**, we evaluate by interpolation the maximum value c_{max} or, the minimum value, respectively c_{min} of the winding ratio compatible with the problem data.

2. The measure unit-less parameter is calculated:

$$A = L_n \sqrt{\frac{T_{kn}}{F_n}}$$

3. Spring flexibility requirement **φ**, measured in mm/N, or stiffness **R$_g$**, measured in N/mm, is calculated using one of the formulas:

$$\phi = \frac{\Delta L}{F_n - F_1}$$

$$R_g = \frac{F_n - F_1}{\Delta L}$$

4. The measure unitless parameter **η** is calculated using one of the two following formulas, depending on whether you have the flexibility or the spring stiffness:

$$\eta = G\phi \sqrt{\frac{F_n}{T_{kn}}}$$

$$\eta = \frac{G}{R_g} \sqrt{\frac{F_n}{T_{kn}}}$$

5. You enter into the diagram shown in **Figure 5** with the value of the parameter **A**, calculated in point 2, and we obtain, directly by interpolation, at the value of the measure unitless parameter **η** calculated in step 4, the winding ratio **c**.

6. If the value found of **c** is less than or equal to c_{max}, evaluated in point 1, this is the solution of the problem in case the maximum winding external diameter of the spring has been assigned. If the value found of **c** it is greater than or equal to c_{min}, evaluated in point 1, this is the solution of the problem in case the internal minimum winding diameter of the spring has been assigned, and If, alternatively in the two cases $c > c_{max}$ or $c < c_{min}$, there are no compatible solutions with the assigned starting data.

7. In case the value found of **c** constitutes a valid solution to the problem, **Annex I** gives us, in correspondence to it, the value of W_e or the value of W_i of the torsion stress per unit of pressure exerted on the basis of the spring, respectively in the two cases in which the maximum external diameter or the minimum internal winding diameter has been assigned.

8. Once known the compatible value of W_e or of W_i, the outer winding diameter is deduced D_e, or, respectively, the internal winding diameter D_i using the formulas:

$$D_e = \sqrt{\frac{4w_e F_n}{\pi T_{kn}}}$$

$$D_i = \sqrt{\frac{4w_i F_n}{\pi T_{kn}}}$$

9. The value to be assigned to the diameter **d**, of the wire or bar with which the spring must be constructed, is obtained using the formulas:

$$d = \frac{D_e}{c+1}$$

$$d = \frac{D_i}{c-1}$$

10. The number of active coils of the spring is calculated using one of the two formulas:

$$i = \frac{Gd\phi}{8c^3}$$

$$i = \frac{Gd}{8c^3 R_g}$$

11. Due to the limitation imposed in § 14.4 about the shape of the ends, the total number of coils of the spring is expressed by the single formula:

$$i_t = i + 2$$

12. The diameter of the wire or bar calculated, in § 14.4.2 point 9, only exceptionally coincides with one of the unified or commercially available value. The total number of coils, calculated in § 14.4.2 point 11, does not usually meet the condition indicated in § 7.1 of the standard. It is therefore necessary to round off the calculated diameter **d** in **d'** and change the total number of coils calculated it in i_t' and, consequently, the number of active coils **i** in **i'**. The variation must be kept to the minimum necessary.

13. Once the constructive values **d'**, i_t' and **i'** have been chosen, the value of the winding ratio **c'** is calculated by the required flexibility or stiffness, using the formulas:

$$c' = \frac{1}{2}\sqrt[3]{\frac{Gd'\phi}{i'}}$$

$$c' = \frac{1}{2} \sqrt[3]{\frac{Gd'}{i'R_g}}$$

14. From **Annex I**, in correspondence of **c'** the value of the stress correction factor **k'** is obtained and we calculate the torsion stress T_{kn} corresponding to the load F_n using the formula:

$$T'_{kn} = \frac{8}{\pi} \frac{c'k'}{d'^2} F_n$$

15. if the stress T_{kn}' is considered acceptable, the control calculation is completed evaluating the spring external winding diameter D_e' and the spring internal winding diameter D_i', its block length L_0' and its minimum inter-coils working length S', using the formulas:

$$D_e' = d'(c' + 1)$$
$$D' = d'(C' - 1)$$
$$L_0' = d'(i' + 1,5)$$
$$S' = L_n - L_b$$

16. If the tension T_{kn}' is too high, or if any of the quantities calculated in § 14.4.2 point 15 is not acceptable, different values are chosen **d''**, i_t'' and **i''** and the control calculation is repeated.

17. The spring free length L_0, measured in mm, can be evaluated using one of the formulas:

$$L_0 = L_n + F_n \varphi$$
$$L_0 = L_n + F_n / R_g$$

18. In the diagram shown in **Figure 5**, in addition to the curves corresponding to constant values of the factor **η**, we have also plotted the curves corresponding to constant values of the number of active coils **i**. Due to their low approximation they are not used directly in the calculation, but allow us an immediate rough evaluation of the active coils

number that will have the spring we are drawing.

14.4.3 Diagram of the calculation procedure in the case where the development length of the wire or the bar, or the mass of the spring, is prefixed

The starting data available for the sizing calculation are those listed in § 14.4.2, with the exception of the length L_n of the spring under the load F_n. In place of these values, they are assigned, alternatively:

- The length of development Σ of the wire or bar, with which the spring must be made;
- The mass Q of the spring, or better the mass of the wire or bar, *before grinding the ends*.

1. The Length of development Σ of the wire or bar can be measured in mm, in first approximation using the formula:

$$\Sigma = \pi D(i+2)$$

2. The mass Q of the wire or bar measured in kg, can be calculated as a first approximation using the formula:

$$Q = \frac{\pi}{4}\rho\Sigma d^2 10^{-6}$$

being indicated with ρ the volumetric mass of the wire or bar constituting the spring measured in kg/dm^3.

3. In place of the measure unitless parameter A, valid for the case in which the spring length L_n under the load F_n has been assigned, we use the two measure unitless parameters B and C.

The measure unitless parameter B, valid for the case in which the development length of the wire or bar Σ has been assigned, is:

$$B = \Sigma\sqrt{\frac{T_{Kn}}{F_n}}$$

The diagram shown in **Figure 6** provides the representative curves of the parameter **B** according to the winding ratio **c**, for different parameter values η defined in § 14.4.2 point

3.

4. The measure unitless parameter **C**, valid for the case where the mass of the thread or of the bar **Q** was assigned, is:

$$C = \frac{1}{2}\frac{Q}{\rho}\left|\frac{T_{kn}}{F_n}\right|^{3/2}$$

The diagram shown in **Figure 7** provides the representative curves of the parameter **C** according to the winding ratio **c** for different values of the parameter η defined in § 14.4.2 point 3.

5. The calculation procedure complies fully with that given in § 14.4.2, with the only difference that, instead of the diagram shown in Figure 5, the diagram shown in **Figure 6** must be used, in the case where the development length Σ, or the diagram shown in **Figure 7**, if the spring mass **Q** has been assigned.

6. We have to note that, given the trend of the curves **B = f (C)** of the diagram of **Figure 6** and of the curves **C = f (c)** of the diagram of **Figure 7**, in some cases it's possible to find two solutions compatible with the assigned starting data. It is in the faculty of the designer to choose the most convenient solution between the two, based on the specific conditions of use of the spring.

7. Both In the diagram shown in **Figure 6** and in the diagram shown in **Figure 7**, in addition to the curves corresponding to constant values of the factor η, we have also plotted the curves corresponding to constant values of the number of active coils. For these curves, the observation made in § 14.4.2 point 10 applies to the corresponding curves of the diagram shown in **Figure 5**

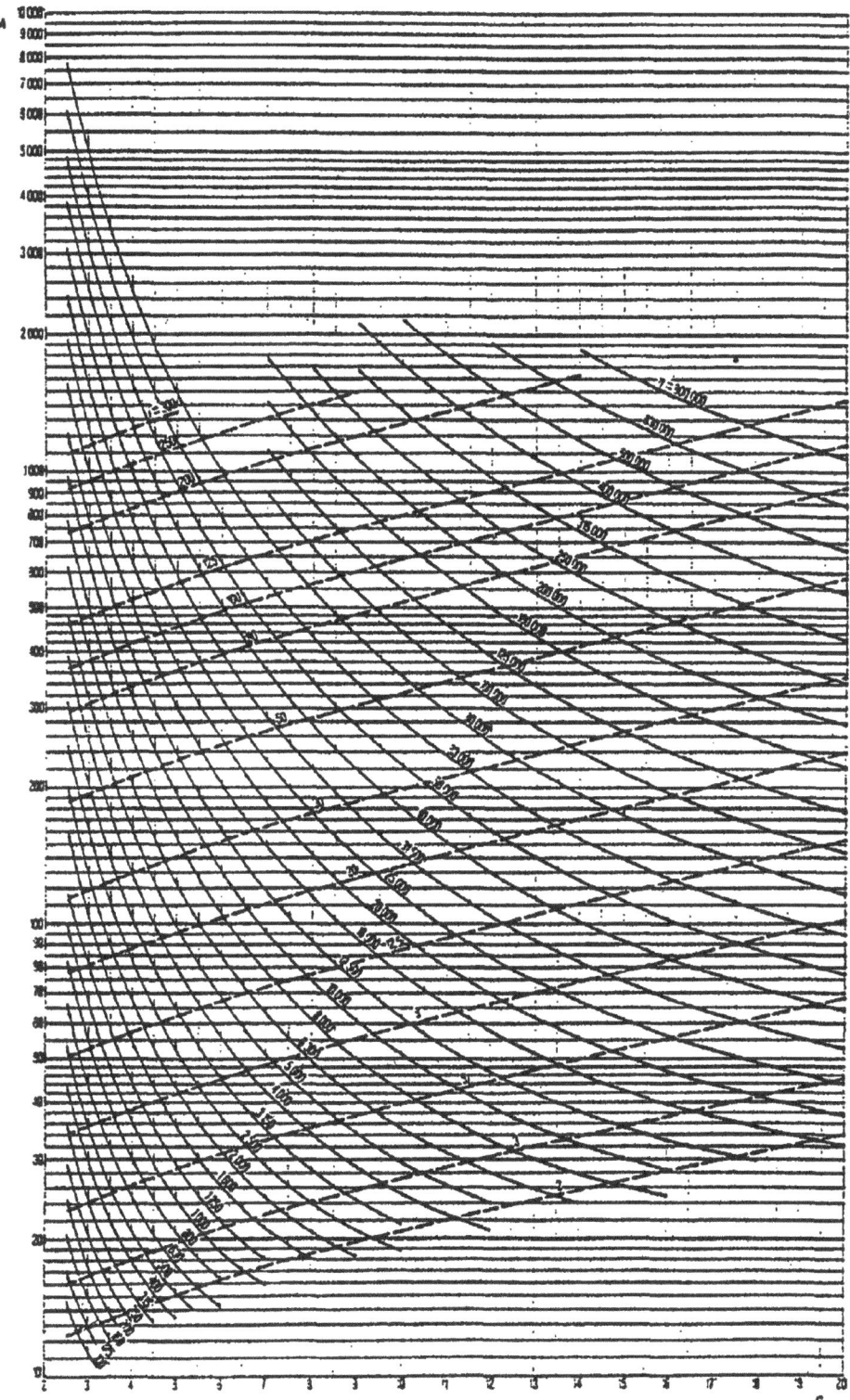

Figure-5 Parameter A according to the winding ratio c

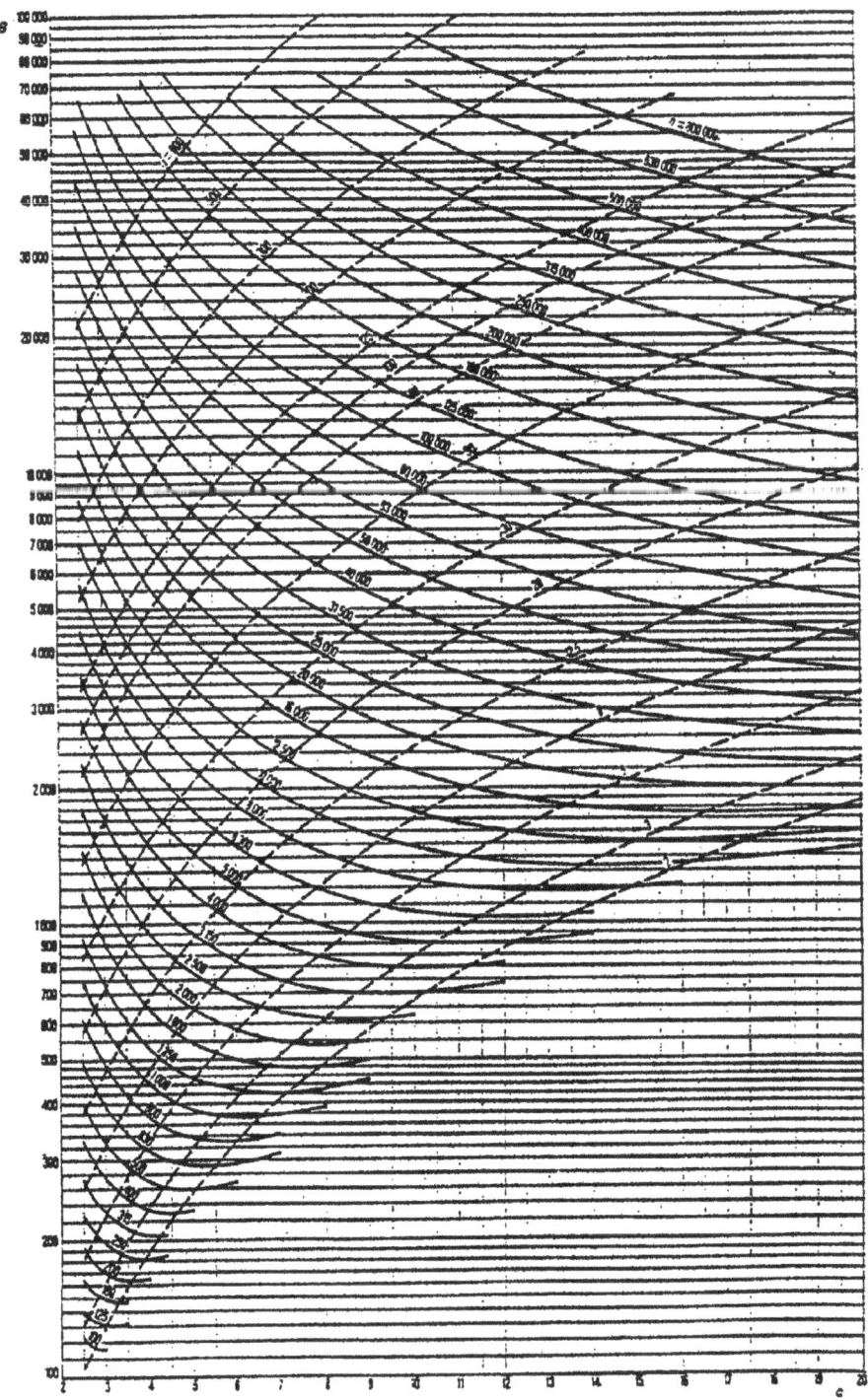

Figure-6 Parameter B according to the winding ratio c

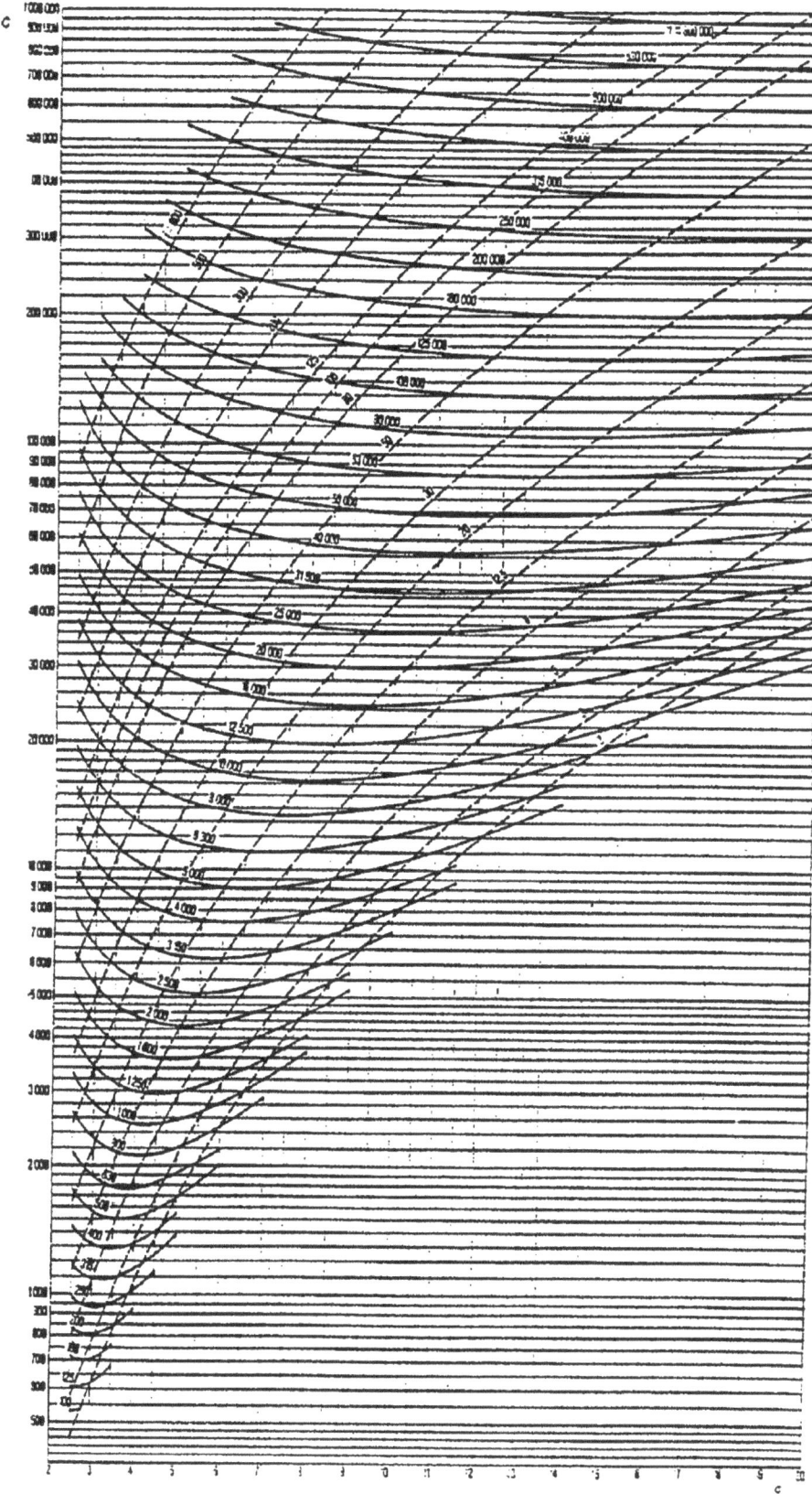

Figure-7 Parameter C according to the winding ratio c

[1]

To obtain the value of S relative to a spring, we proceed as indicated in the following example:

d = 4.30, D = 24, c = 5.58, i = 6.5

for c = 6 s = 0.109

for c = 5 s = 0.100

Δs = 0.009

0.009 x (58/100) = 0.00522

for c = 5 s = 0.100

for c = 5.58 s = 0.10522

Therefore it results: S = 0.10522 x 6.5 (4.3 + 0.5) = 3.28

Correction factor k of the torsion stress and torsion stress per unit of pressure W, We, Wi

C	k	W	We	Wi
3,00	1,580	85,3	151,7	37,9
3,05	1,567	88,9	156,8	40,2
3,10	1,555	92,7	162,1	42,5
3,15	1,544	96,5	167,5	45,0
3,20	1,533	100,5	173,1	47,5
3,25	1,523	104,5	178,8	50,1
3,30	1,513	108,7	184,6	52,8
3,35	1,503	113,0	190,5	55,6
3,40	1,493	117,4	196,6	58,5
3,45	1,484	121,9	202,8	61,5
3,50	1,475	126,5	209,2	64,6
3,55	1,467	131,3	215,7	67,7
3,60	1,459	136,2	222,3	71,0
3,65	1,451	141,2	229,1	74,4
3,70	1,444	146,3	236,0	77,9
3,75	1,437	151,5	243,1	81,5
3,80	1,430	156,9	250,3	85,2
3,85	1,423	162,4	257,7	89,0
3,90	1,416	168,0	265,2	92,9
3,95	1,410	173,8	272,9	96,9
4,00	1,404	179,7	280,7	101,1

4,05	1,398	185,7	288,7	105,3
4,10	1,392	191,9	296,9	109,7
4,15	1,386	198,2	305,2	114,2
4,20	1,381	204,6	313,6	118,8
4,25	1,375	211,2	322,2	123,5
4,30	1,370	217,9	331,0	128,3
4,35	1,365	224,8	340.0	133,3
4,40	1,360	231,8	349,1	138,4
4,45	1,356	238,9	358,4	143,6
4,50	1,351	246,2	367,8	148,9
4,55	1,346	253,7	377,7	154,4
4,60	1,342	261,3	387,2	160,0
4,65	1,338	269,0	397,1	165,7
4,70	1,334	276,9	407,3	171,6
4,75	1,330	285,0	417,6	177,6
4,80	1,326	293,2	428,1	183,7
4,85	1,322	301,6	438,7	190,0
4,90	1,318	310,1	449,6	196,4
4,95	1,314	318,8	460,6	203,0
5,00	1,311	327,6	471,8	209,7
5,10	1,304	345,8	494,7	223,5
5,20	1,297	364,7	518,4	237,9
5,30	1,290	384,2	542,9	252,9
5,40	1,284	404,5	568,2	268,5
5,50	1,278	425,4	594,2	284,8
5,60	1,273	447,1	621,0	301,7
5,70	1,267	469,4	648,6	319,2
5,80	1,262	492,6	677,1	337,4
5,90	1,257	516,4	706,3	356,2
6,00	1,253	541,1	736,5	375,7

6,10	1,248	566,5	767,4	396,0
6,20	1,243	592,7	799,3	416,9
6,30	1,239	619,7	832,0	438,6
6,40	1,235	647,5	865,6	461,0
6,50	1,231	676,1	900,2	484,1
6,60	1,227	705,6	935,6	508,0
6,70	1,223	735,9	971,9	532,6
6,80	1,220	767,1	1009	558,0
6,90	1,216	799,1	1048	584,3
7,00	1,213	832,0	1087	611,3
7,10	1,210	865,8	1127	639,1
7,20	1,206	900,6	1168	667,8
7,30	1,203	936,8	1210	697,3
7,40	1,200	972,8	1253	727,6
7,50	1,197	1010	1298	758,8
7,60	1,195	1049	1343	790,9
7,70	1,192	1088	1389	823,9
7,80	1,189	1129	1437	857,8
7,90	1,187	1170	1485	892,6
8,00	1,184	1212	1534	928,3
8,10	1,182	1256	1585	964,9
8,20	1,179	1300	1637	1002
8,30	1,177	1346	1690	1041
8,40	1,175	1392	1744	1081
8,50	1,172	1440	1799	1121
8,60	1,170	1489	1855	1163
8,70	1,168	1538	1912	1205
8,80	1,166	1589	1971	1249
8,90	1,164	1641	2031	1293

9,00	1,162	1694	2092	1339
9,10	1,160	1749	2154	1385
9,20	1,158	1804	2217	1433
9,30	1,156	1860	2282	1482
9,40	1,155	1918	2348	1532
9,50	1,153	1977	2415	1583
9,60	1,151	2037	2484	1635
9,70	1,150	2098	2553	1688
9,80	1,148	2161	2624	1742
9,90	1,146	2225	2697	1798
10,00	1,145	2290	2770	1855
10,20	1,142	2432	2922	1972
10,40	1,139	2562	3079	2093
10,60	1,136	2706	3241	2220
10,80	1,133	2856	3409	2351
11,00	1,131	3010	3583	2488
11,20	1,128	3171	3762	2639
11,40	1,126	3337	3948	2777
11,60	1,124	3508	4139	2929
11,80	1,122	3686	4337	3087
12,00	1,119	3869	4540	3251
12,20	1,117	4058	4750	3420
12,40	1,115	4253	4967	3595
12,60	1,113	4454	5190	3776
12,80	1,112	4662	5419	3962
13,00	1,110	4876	5656	4155
13,50	1,106	5440	6276	4664
14,00	1,102	6046	6940	5213
14,50	1,098	6695	7650	5803
15,00	1,095	7388	8406	6436

ANNEX I

15,50	1,091	8128	9211	7113
16,00	1,088	8916	10066	7837
16,50	1,086	9754	10972	8607
17,00	1,083	10642	11931	9427
17,50	1,081	11583	12944	10297
18,00	1,078	12577	14013	11218
18,50	1,076	13627	15140	12194
19,00	1,074	14734	16325	13224
19,50	1,072	15899	17571	14310
20,00	1,070	17124	18879	15434

www.ingramcontent.com/pod-product-compliance
Lightning Source LLC
Chambersburg PA
CBHW081639220526
45468CB00009B/2498